THE POETRY OF DARMSTADTIUM

The Poetry of Darmstadtium

Walter the Educator

Silent King Books

SILENT KING BOOKS

SKB

Copyright © 2024 by Walter the Educator

All rights reserved. No part of this book may be reproduced in any manner whatsoever without written permission except in the case of brief quotations embodied in critical articles and reviews.

First Printing, 2024

Disclaimer
This book is a literary work; poems are not about specific persons, locations, situations, and/or circumstances unless mentioned in a historical context. This book is for entertainment and informational purposes only. The author and publisher offer this information without warranties expressed or implied. No matter the grounds, neither the author nor the publisher will be accountable for any losses, injuries, or other damages caused by the reader's use of this book. The use of this book acknowledges an understanding and acceptance of this disclaimer.

"Earning a degree in chemistry changed my life!"
- Walter the Educator

dedicated to all the chemistry lovers, like myself, across the world

DARMSTADTIUM

In the abyss of the periodic table's vast expanse,

DARMSTADTIUM

Where elements dance in atomic trance,

DARMSTADTIUM

Darmstadtium emerges, a fleeting glance,

DARMSTADTIUM

A shimmering specter, a cosmic advance.

DARMSTADTIUM

Born in the depths of a nuclear forge,

DARMSTADTIUM

Where fusion and fusion converge,

DARMSTADTIUM

It springs to life with atomic surge,

DARMSTADTIUM

A fleeting moment, a celestial dirge.

DARMSTADTIUM

Named for the city where discovery reigned,

DARMSTADTIUM

In laboratories where minds are trained,

DARMSTADTIUM

Darmstadtium's essence, so briefly contained,

DARMSTADTIUM

In fleeting moments, its presence sustained.

DARMSTADTIUM

With atomic number one hundred and ten,

DARMSTADTIUM

It joins the ranks of the rarest, the zen,

DARMSTADTIUM

A transient being, unseen by most men,

DARMSTADTIUM

Yet in its existence, a tale to transcend.

DARMSTADTIUM

In the heart of a collider's collision ballet,

DARMSTADTIUM

Darmstadtium particles come out to play,

DARMSTADTIUM

For fractions of moments, they hold sway,

DARMSTADTIUM

Then vanish into the cosmic array.

DARMSTADTIUM

With nuclei unstable, it yearns to decay,

DARMSTADTIUM

Releasing its energy, fading away,

DARMSTADTIUM

A fleeting existence, a cosmic display,

DARMSTADTIUM

In the grand symphony of atoms at play.

DARMSTADTIUM

Yet in its brief lifespan, it leaves a mark,

DARMSTADTIUM

A testament to human ingenuity's spark,

DARMSTADTIUM

For in the quest to unravel the dark,

DARMSTADTIUM

Darmstadtium shines, a celestial landmark.

DARMSTADTIUM

Its properties mysterious, its behavior unique,

DARMSTADTIUM

In the realm of the elements, it's the meek,

DARMSTADTIUM

Yet in its rarity, a story to speak,

DARMSTADTIUM

Of scientific triumphs, of knowledge to seek.

DARMSTADTIUM

So let us marvel at Darmstadtium's grace,

DARMSTADTIUM

In the cosmic dance of time and space,

DARMSTADTIUM

A fleeting presence, a celestial embrace,

DARMSTADTIUM

In the tapestry of elements, a shimmering trace.

DARMSTADTIUM

As we ponder its mysteries, let us aspire,

DARMSTADTIUM

To reach for the stars, to aim ever higher,

DARMSTADTIUM

For in the quest for knowledge, we never tire,

DARMSTADTIUM

And Darmstadtium's legacy will never expire.

DARMSTADTIUM

ABOUT THE CREATOR

Walter the Educator is one of the pseudonyms for Walter Anderson. Formally educated in Chemistry, Business, and Education, he is an educator, an author, a diverse entrepreneur, and he is the son of a disabled war veteran. "Walter the Educator" shares his time between educating and creating. He holds interests and owns several creative projects that entertain, enlighten, enhance, and educate, hoping to inspire and motivate you.

Follow, find new works, and stay up to date
with Walter the Educator™
at WaltertheEducator.com

www.ingramcontent.com/pod-product-compliance
Lightning Source LLC
LaVergne TN
LVHW010412070526
838199LV00064B/5272